North Korea Rocket Launch and Hydrogen Test

It's Implication

By

Buchi Nwadiuto

DEDICATION

This book is dedicated to you

PREFACE

North Korea launched a satellite into space. Its state media reported, triggering a wave of international condemnation and prompting strong reaction from an emergency meeting of the U.N. Security Council.

Though North Korea said the launch was for scientific and "peaceful purposes," it is being widely viewed by other nations as a front to test a ballistic missile, especially coming on the heels of North Korea's purported hydrogen bomb test last month. Pyongyang carried out both acts in defiance of international sanctions.

At an emergency meeting, members of the U.N. Security Council "strongly condemned" the launch and reaffirmed that "a clear threat to international peace and security continues to exist, especially in the context of the nuclear test." Security Council members have previously threatened "further significant measures" if there was another North Korean launch and now will "adopt expeditiously a new Security Council resolution with such measures in response to these dangerous and serious violations," according to a statement read by Venezuela's ambassador to the United Nations after the meeting.

U.N. Secretary-General Ban Ki-moon said the launch is "deeply deplorable" and in violation of Security Council Resolution.

Table of Contents

CHAPTER ONE
HISTORY OF THE NUCLEAR PROGRAM

The nuclear program can be traced back to about 1962, when North Korea committed itself to what it called "all-fortressization", which was the beginning of the hyper-militarized North Korea of today. In 1963 North Korea asked the Soviet Union for help in developing nuclear weapons, but was refused. The Soviet Union agreed to help North Korea develop a peaceful nuclear energy program, including the training of nuclear scientists. Later, China, after its nuclear tests, similarly rejected North Korean requests for help with developing nuclear weapons.

Soviet specialists took part in the construction of the Yongbyon Nuclear Scientific Research Center and began construction of an IRT-2000 research reactor in 1963, which became operational in 1965 and was upgraded to 8 MW in 1974. In 1979 North Korea indigenously began to build in Yongbyon a second research reactor, an ore processing plant and a fuel rod fabrication plant.

North Korea's nuclear weapons program dates back to the 1980s. Focusing on practical uses of nuclear energy and the completion of a nuclear weapon development system, North Korea began to operate facilities for uranium fabrication and conversion, and conducted high-explosive detonation tests. In 1985 North Korea ratified the NPT, but did not conclude the required safeguards agreement with the IAEA until 1992. In early 1993, while verifying North Korea's initial declaration, the IAEA concluded that there was strong evidence this declaration was incomplete. When North Korea refused the requested special inspection, the IAEA reported its non-compliance to the UN Security Council. In 1993, North Korea announced its withdrawal from the NPT, but suspended that withdrawal before it took effect.

Under the 1994 Agreed Framework, the U.S. government agreed to facilitate the supply of two light water reactors to North Korea in exchange for North Korean disarmament. Such reactors are considered "more proliferation-resistant than North Korea's graphite-moderated reactors", but not "proliferation proof". Implementation of the Agreed Framework floundered, and in 2002 the Agreed Framework fell apart, with each side blaming the other for its failure. By 2002, Pakistan had admitted that North Korea had gained access to Pakistan's nuclear technology in the late 1990s. Based on evidence from Pakistan, Libya, and multiple confessions from North Korea itself, the United States accused North Korea of non-compliance and halted oil shipments; North Korea later claimed its public confession of guilt had been deliberately misconstrued. By the end of 2002, the Agreed Framework was officially dead.

In 2003, North Korea again announced its withdrawal from the Nuclear Proliferation Treaty. In 2005, it admitted to having nuclear weapons but vowed to close the nuclear program.

On March 17, 2007, North Korea told delegates at international nuclear talks that it is preparing to shut down its main nuclear facility. The agreement was reached following a series of six-party talks, involving North Korea, South Korea, China, Russia, Japan, and the United States begun in 2003. According to the agreement, a list of its nuclear programs will be submitted and the nuclear facility will be disabled in exchange for fuel aid and normalization talks with the United States and Japan. This was delayed from April due to a dispute with the United States over Banco Delta Asia, but on July 14, International Atomic Energy Agency inspectors confirmed the shutdown of North Korea's Yongbyon nuclear reactor and consequently North Korea began to receive aid. This agreement fell apart in 2009, following a North Korean missile test.

In February 2012, North Korea announced that it would suspend uranium enrichment at the Yongbyon Nuclear Scientific Research Center and not conduct any further tests of nuclear weapons while productive negotiations involving the United States continue. This agreement included a moratorium on long-range missiles tests. Additionally, North Korea agreed to allow IAEA inspectors to monitor operations at Yongbyon. The United States reaffirmed that it had no hostile intent toward the DPRK and was prepared to improve bilateral relationships, and agreed to ship humanitarian food aid to North Korea. The United States called the move "important, if limited", but said it would proceed cautiously and that talks would resume only after North Korea made steps toward fulfilling its promise. However, after North Korea conducted a long-range missile test in April 2012, the United States decided not to proceed with the food aid.

CHAPTER TWO
NUCLEAR WEAPON

The Korean Central News Agency claims that "The Bush administration's DPRK policy that stemmed from its ignorance of the DPRK resulted in making the DPRK a nuclear weapons state." North Korea had been suspected of maintaining a clandestine nuclear weapons development program since the early 1980s, when it constructed a plutonium-producing Magnox nuclear reactor at Yongbyon. Various diplomatic means had been used by the international community to attempt to limit North Korea's nuclear program to peaceful power generation and to encourage North Korea to participate in international treaties. During the 13th World Festival of Youth and Students held in the DPRK in 1989, South Korean activist and "Flower of Reunification" Lim Su-kyung implied that the DPRK was not seeking nuclear weapons, saying: "The slogan 'Let us build a new world free from nuclear weapons!' will not be materialized by words alone. I'd like you to resolutely struggle against the anti-reunification forces, and give us support and encouragement. I, too, want to live in a country free from nuclear weapons; in my own land, and not infested with foreign forces and foreign army troops."

In May 1992, North Korea's first inspection by the International Atomic Energy Agency (IAEA) uncovered discrepancies suggesting that North Korea had reprocessed more plutonium than declared. IAEA requested access to additional information and access to two nuclear waste sites at Yongbyon. North Korea rejected the IAEA request and announced on March 12, 1993, an intention to withdraw from the NPT.

In 1994, North Korea pledged, under the "Agreed Framework" with the United States, to freeze its plutonium programs and dismantle all its nuclear weapons programs in return for several kinds of assistance, including construction of two modern nuclear power plants powered by light-water reactors.

By 2002, the United States believed that North Korea was pursuing both uranium enrichment technology and plutonium reprocessing technologies in defiance of the Agreed Framework. North Korea reportedly told American diplomats in private that they were in possession of nuclear weapons, citing American failures to uphold their own end of the "Agreed Framework" as a motivating force. North Korea later "clarified" that it did not possess weapons yet, but that it had "a right" to possess them, despite the Agreed Framework. In late 2002 and early 2003, North Korea began to take steps to eject International Atomic Energy Agency inspectors while re-routing spent fuel rods for plutonium reprocessing for weapons purposes. As late as the end of 2003, North Korea claimed that it would freeze its nuclear program in exchange for additional American concessions, but a final agreement was not reached.

North Korea withdrew from the Nuclear Non-Proliferation Treaty in 2003.

On October 9, 2006, North Korea demonstrated its nuclear capabilities with its first underground nuclear test, detonating a plutonium based device and the estimated yield was 0.2–1 kiloton.[11] The test was conducted at P'unggye-yok, and U.S. intelligence officials later announced that analysis of radioactive debris in air samples collected a few days after the test confirmed that the blast had taken place. The United Nations Security Council condemned the test in Resolution 1874.

On January 6, 2007, the North Korean government further confirmed that it had nuclear weapons.

In February 2007, following the six-party talks disarmament process, Pyongyang agreed to shut down its main nuclear reactor. On October 8, 2008, IAEA inspectors were forbidden by the North Korean government to conduct further inspections of the site.

On April 25, 2009, the North Korean government announced that the country's nuclear facilities had been reactivated, and that spent fuel reprocessing for arms-grade plutonium has been restored.

On May 25, 2009, North Korea conducted its second underground nuclear test. The U.S. Geological Survey calculated its origin in proximity of the site of the first nuclear test. The test was more powerful than the previous test, estimated at 2 to 7 kilotons. The same day, a successful short range missile test was also conducted.

2013

On February 12, monitors in Asia picked up unusual seismic activity at a North Korean facility at 11:57am (02:57 GMT), later determined to be an artificial quake with an initial magnitude 4.9 (later revised to 5.1). The Korean Central News agency subsequently said that the country had detonated a miniaturized nuclear device with "greater explosive force" in an underground test. According to the Korea Institute of Geosciences and Mineral Resources, the estimated yield was 7.7–7.8 kilotons.

2015

In December 2015, Kim Jong-un suggested that the country had the capacity to launch a hydrogen bomb, a device of considerably more power than conventional atomic bombs used in previous tests. The remark was met with skepticism from the White House and South Korean officials.

2016

On January 6, after reports of a magnitude 5.1 earthquake originating in northeast North Korea at 10:00:01 UTC+08:30, the country's regime released statements that it had successfully tested a hydrogen bomb. Whether this was in fact a hydrogen bomb has yet to be proven. Experts have cast doubt on this claim. A South Korean spy expert suggested that it may have been an atomic bomb and not a hydrogen bomb. Experts in several countries, including South Korea have expressed doubts about the claimed technology because of the relatively small size of the explosion. Senior Defense Analyst Bruce W. Bennett of research organization RAND told the BBC that "... Kim Jong-un is either lying, saying they did a hydrogen test when they didn't, they just used a little bit more efficient fission weapon – or the hydrogen part of the test really didn't work very well or the fission part didn't work very well."

Nations across the world, as well as NATO and the United Nations, have spoken out against the testing as destabilizing, as a danger to international security and as a breach of UN Security Council resolutions. China, one of North Korea's only allies, also denounced the test.

CHAPTER THREE
CHEMICAL AND BIOLOGICAL WEAPONS

North Korea acceded to the Biological Weapons Convention in 1987 and the Geneva Protocol on January 4, 1989, but has not signed the Chemical Weapons Convention.

The U.S. Department of Defense believes North Korea probably has a chemical weapons program and is likely to possess a stockpile of weapons. The United States believes that North Korea maintains a biological weapons capability and infrastructure, and has the munitions production capacity to deploy biological agents if it chose to do so.

North Korea reportedly acquired the technology necessary to produce tabun and mustard gas as early as the 1950s. The United States estimates North Korea's likely stockpile of chemical weaponry from at least a few hundred tons, to at most a few thousand tons.

In 2009 the International Crisis Group reported that the consensus expert view was that North Korea had a stockpile of about 2,500 to 5,000 tons of chemical weapons, including mustard gas, sarin (GB) and other nerve agents. The South Korean government also estimated the stockpile as about 2,500 to 5,000 tons in 2010.

North Korea may have also begun the production of binary agents. Binary agents are toxic only when the two chemicals (normally physically separated) are combined. By creating binary agents, North Korea can increase their safety when handling hazardous material. North Korean military units conduct regular nuclear, biological, and chemical (NBC) training exercises in a chemical environment. North Korean chemical and biological warfare units are equipped with decontamination and detection equipment. In 2010, the Omaha World-Herald reported that North Korea has chemical weapons which could cause millions of casualties in South Korea, where gas masks are only provided to the military and top government officials.

On June 6, 2015, a North Korean defector to Finland who is working in China claims to have 15 gigabytes of electronic evidence that he claims documents how the country is testing chemical and biological agents on its own citizens. The same day, photo releases of Kim Jong-un visiting the Pyongyang Bio-technical Institute were scrutinized by experts such as Melissa Hanham of the James Martin Center for Nonproliferation Studies, who claims that this factory is an anthrax-producing factory. However, an official spokesperson for the National Defense Commission denied the allegations on the Korean Central News Agency, challenging the US Congress to inspect the Institute, saying: "Come here right now, with all the 535 members of the House of Representatives and the Senate as well as the imbecile secretaries and deputy secretaries of the government who have made their voices hoarse screaming for new sanctions. Then they can behold the awe-inspiring sight of the Pyongyang Bio-technical Institute."

CHAPTER FOUR
DELIVERY SYSTEM

There is evidence that North Korea has been able to miniaturize a nuclear warhead for use on a ballistic missile. Re-entry technology to protect the warheads en route to their targets is lacking. The April 2012 display of missiles purporting to be ICBMs were declared fakes by Western analysts, and indicated North Korea was a long way from having a credible ICBM. In December 2012, North Korea placed a satellite into orbit for the first time.

Successfully tested

KN-1 – a short-range anti-ship cruise missile. Its range is estimated to be around 160 kilometers, and is most probably an improved version of the Soviet Termit missile (NATO codename "Styx").

KN-2 Toksa – a short-range, solid-fueled, highly accurate mobile missile, modified copy of the Soviet OTR-21. Unknown number in service apparently deployed either in the late 1990s or early 2000s (decade).

Hwasong-5 – initial Scud modification. Road-mobile, liquid-fueled missile, with an estimated range of 330 km. It has been tested successfully. It is believed that North Korea has deployed some 150–200 such missiles on mobile launchers.

Hwasong-6 – later Scud modification. Similar to the Hwasong-5, yet with an increased range (550–700 km) and a smaller warhead (600–750 kg). Apparently this is the most widely deployed North Korean missile, with at least 400 missiles in use.

Nodong-1 – larger and more advanced Scud modification. Liquid-fueled, road-mobile missile with a 650 kg warhead. First production variants had inertial guidance; later variants featured GPS guidance, which improves CEP accuracy to 190–250 m. Range is estimated to be between 1,300 and 1,600 km.

Taepodong-1 – three-stage technology demonstrator test bed. First stage was adapted from a Rodong-1. Second stage was adapted from a Hwasong-6. A satellite-delivery launch was attempted in 1998. The satellite failed, but the first two stages apparently functioned adequately. According to some analysts, the Taepodong-1, if developed into an ICBM platform, could have a range of nearly 6,000 km with a third stage and a payload of less than 100 kg. The U.S. Defense Intelligence Agency estimates that the Taepodong-1 was a test-bed, not intended or usable as a weapon. The US National Air and Space Intelligence Center made a similar assessment.

Untested / failed

Taepodong-2 – Three-stage technology demonstrator. First test occurred in 2006, when the missile failed 40 seconds after launch. Estimates of the range vary widely – from 4,500 to 10,000 kilometers (most estimates put the range at about 6,700 km). As of 2013, the Taepodong-2 has not yet been deployed.

KN-08 – Road-mobile ICBM. Maximum range >3,400 miles. The US Defense Department estimates at least 6 KN-08 launchers are in deployment.

Musudan – believed to be a modified copy of the Soviet R-27 Zyb SLBM, untested as of 2013. Originally believed to have been tested as the first or second stage of Unha, but debris analysis showed that the Unha used older technology than it is believed the Musudan uses. Also known under the names Nodong-B, Taepodong-X and BM25, predicted to have a range of 2,500–4,000 km assuming R-27 technology is used.[116] A DoD report puts BM25 strength at fewer than 50 launchers.

CHAPTER FIVE
EXPORTS RELATED TO BALLISTIC MISSILE TECHNOLOGY

In April 2009, the United Nations named the Korea Mining and Development Trading Corporation (KOMID) as North Korea's primary arms dealer and main exporter of equipment related to ballistic missiles and conventional weapons. The UN lists KOMID as being based in the Central District, Pyongyang. However, it also has offices in Beijing and sales offices worldwide which facilitate weapons sales and seek new customers for North Korean weapons.

KOMID has sold missile technology to Iran and has done deals for missile related technology with the Taiwanese. KOMID representatives were also involved in a North Korean deal to mass-produce Kornet anti-tank guided missiles for Syria and KOMID has also been responsible for the sale of equipment, including missile technologies, gunboats, and multiple rocket artilleries, worth a total of over $100 million, to Africa, South America, and the Middle East.

North Korea's military has also used a company called Hap Heng to sell weapons overseas. Hap Heng was based in Macau in the 1990s to handle sales of weapons and missile and nuclear technology to nations such as Pakistan and Iran. Pakistan's medium-range ballistic missile, the Ghauri, is considered to be a copy of North Korea's Rodong 1. In 1999, intelligence sources claim that North Korea had sold missile components to Iran. Listed directors of Hap Heng include Kim Song in and Ko Myong Hun. Ko Myong Hun is now a listed diplomat in Beijing and may be involved in the work of KOMID.

A UN sanctions committee report stated that North Korea operates an international smuggling network for nuclear and ballistic missile technology, including to Myanmar (Burma), Syria, and Iran.

CHAPTER SIX
THE 1994 CRISIS AND AGREED FRAMEWORK

North Korea finally signed an IAEA safeguards agreement on 30 January 1992, and the Supreme People's Assembly ratified the agreement on 9 April 1992. Under the terms of the agreement, North Korea provided an "initial declaration" of its nuclear facilities and materials, and provided access for IAEA inspectors to verify the completeness and correctness of its initial declaration. Six rounds of inspections began in May 1992 and concluded in February 1993. Pyongyang's initial declaration included a small plutonium sample (less than 100 grams), which North Korean officials said was reprocessed from damaged spent fuel rods that were removed from the 5MW reactor in Yongbyon-kun. However, IAEA analysis indicated that Korean technicians had reprocessed plutonium on three occasions—in 1989, 1990, and 1991. When the Agency requested access to two suspect nuclear waste sites, North Korea declared them to be military sites and therefore off-limits.

After the IAEA was denied access to North Korea's suspect waste sites in early 1993, the Agency asked the United Nations Security Council (UNSC) to authorize special ad hoc inspections. In reaction, North Korea announced its intention to withdraw from the NPT on 12 March 1993. Under the terms of the treaty, a state's withdrawal does not take effect until 90 days after it has given notice. Following intense bilateral negotiations with the United States, North Korea announced it was suspending its withdrawal from the NPT one day before the withdrawal was to take effect. Pyongyang agreed to suspend its withdrawal while talks continued with Washington, but claimed to have a special status in regard to its nuclear safeguards commitments. Under this special status, North Korea agreed to allow the continuity of safeguards on its present activities, but refused to allow inspections that could verify past nuclear activities.

As talks with the United States over North Korea's return to the NPT dragged on, North Korea continued to operate its 5MW reactor in Yongbyon. On 14 May 1994, Korean technicians began removing the reactor's spent fuel rods without the supervision of IAEA inspectors. This action worsened the emerging crisis because the random placement of the spent fuel rods in a temporary storage pond compromised the IAEA's capacity to reconstruct the operational history of the reactor, which could have been used in efforts to account for the discrepancies in Pyongyang's reported plutonium reprocessing. U.S. President Bill Clinton's administration announced that it would ask the UNSC to impose economic sanctions; Pyongyang responded that it would consider economic sanctions "an act of war."

The crisis was defused in June 1994 when former U.S. President Jimmy Carter traveled to Pyongyang to meet with Kim II Sung. Carter announced from Pyongyang that Kim had accepted the broad outline of a deal that was later finalized as the Agreed Framework in October 1994. Under the agreement, North Korea agreed to freeze work at its gas-graphite moderated reactors and related facilities, and to allow the IAEA to monitor that freeze. Pyongyang was also required to "consistently take steps to implement the North-South Joint Declaration on the Denuclearization of the Korean Peninsula," and to remain a party to the NPT. In exchange, the United States agreed to lead an international consortium to construct two light water power reactors, and to provide 500,000 tons of heavy fuel oil per year until the first reactor came online with a target date of 2003. Furthermore, the United States was to provide "formal assurances against the threat or use of nuclear weapons by the U.S."

CHAPTER SEVEN
COLLAPSE OF THE AGREED FRAMEWORK AND WITHDRAWAL FROM THE NPT

While the Agreed Framework froze North Korea's plutonium program for almost a decade, neither party was completely satisfied with either the compromise reached or its implementation. The United States was dissatisfied with the postponement of safeguards inspections to verify Pyongyang's past activities, and North Korea was dissatisfied with the delayed construction of the light water power reactors.

After coming to office in 2001, the George W. Bush administration initiated a North Korean policy review, which it completed in early June. The review concluded that the United States should seek "improved implementation of the Agreed Framework, verifiable constraints on North Korea's missile program, a ban on missile exports, and a less threatening North Korean conventional military posture." From Washington's perspective, "improved implementation of the Agreed Framework" meant an acceleration of safeguards inspections, even though the agreement did not require Pyongyang to submit to full safeguards inspections to verify its past activities until a significant portion of the reactor construction was completed, but before the delivery of critical reactor components.

The international community also became concerned that North Korea might have an illicit highly enriched uranium (HEU) program. In the summer of 2002, U.S. intelligence reportedly discovered evidence of transfers of HEU technology and/or materials from Pakistan to North Korea in exchange for ballistic missiles technology. (Later, in early 2004, it was revealed that Pakistani nuclear scientist Dr. A. Q. Khan had sold gas-centrifuge technology to North Korea, Libya and Iran.)

In October 2002, bilateral talks between the United States and North Korea finally resumed when U.S. Assistant Secretary of State for East Asia and Pacific Affairs James Kelly visited Pyongyang. During the visit, Kelly informed First Vice Foreign Minister Kang Sok Chu and Vice Foreign Minister Kim Kye Kwan that Washington was aware of a secret North Korean program to produce HEU. The U.S. State Department claimed that North Korean officials admitted to having such a program during a second day of meetings with Kelly, but North Korea later argued that it had only admitted to having a "plan to produce nuclear weapons," which Pyongyang claimed was part of its right to self-defense.

The United States responded in December 2002 by suspending heavy oil shipments, and North Korea retaliated by lifting the freeze on its nuclear facilities, expelling IAEA inspectors monitoring that freeze, and announcing its withdrawal from the NPT on 10 January 2003. Initially, North Korea claimed it had no intention of producing nuclear weapons, and that the lifting of the nuclear freeze was necessary to generating needed electricity.

CHAPTER EIGHT
INDIGENOUS DEVELOPMENT

In the late 1960s, North Korea expanded its educational and research institutions to support a nuclear program for both civilian and military applications. By the early 1970s, North Korean engineers were using indigenous technology to expand the IRT-2000 research reactor, and Pyongyang had begun to acquire plutonium reprocessing technology from the Soviet Union. In July 1977, North Korea signed a trilateral safeguards agreement with the IAEA and the USSR that brought the IRT-2000 research reactor and a critical assembly in Yongbyon under IAEA safeguards. The Soviets were included in the agreement because they supplied the reactor's fuel.

The early 1980s was a period of significant indigenous expansion, when North Korea constructed uranium milling facilities, fuel rod fabrication complex, and a 5MW(e) nuclear reactor, as well as research and development institutions. Simultaneously, North Korea began experimenting with the high explosives tests required for building the triggering mechanism of a nuclear bomb. By the mid-1980s, Pyongyang had begun constructing a 50MW nuclear reactor in Yongbyon, while also expanding its uranium processing facilities.

Pyongyang also explored the acquisition of light water reactor technology in the early to mid-1980s. This period coincided with the expansion of North Korea's indigenously designed reactor program, which was based on gas-graphite moderated reactors similar in design to the Calder Hall reactors first built in the United Kingdom in the 1950s. Pyongyang agreed to sign the Treaty on the Non-Proliferation of Nuclear Weapons (NPT) as a non-nuclear weapon state in December 1985 in exchange for Soviet assistance constructing four LWRs.

In September 1991, U.S. President George H. W. Bush announced that the United States would withdraw its nuclear weapons from South Korea, and on 18 December 1991, President Roh Tae Woo declared that South Korea was free of nuclear weapons. North Korea and South Korea then signed the Joint Declaration on the Denuclearization of the Korean Peninsula, whereby both sides promised they would "not test, manufacture, produce, receive, possess, store, deploy or use nuclear weapons." The agreement additionally bound the two sides to forgo the possession of "nuclear reprocessing and uranium enrichment facilities." The agreement also provided for a bilateral inspections regime, but the two sides failed to agree on its implementation.

CHAPTER NINE
NEW DEVELOPMENTS

In early 2003, U.S. intelligence detected activities around the Radiochemisty Laboratory, a reprocessing facility in Yongbyon, which indicated that North Korea was probably reprocessing the 8,000 spent fuel rods that had been in a temporary storage pond. In September 2003, a North Korean Foreign Ministry spokesman said that North Korea had completed the reprocessing of this spent fuel—this would have given North Korea enough plutonium for approximately four to six nuclear devices. In January 2004, a delegation of invited U.S. experts confirmed that the canisters in the temporary storage pond were empty.

In April 2003, a multilateral dialogue began in Beijing with the aim of ending Pyongyang's nuclear weapons program. Initially trilateral in format (China, North Korea and the United States), the process expanded to a six-party format with the inclusion of Japan, Russia and South Korea. The first round began in August 2003. Six months later, in February 2004, the second round of talks was held, and a third round followed in June 2004. However, tensions between the parties—particularly the United States and North Korea—caused the talks to stall for more than a year, restarting in July 2005.

While the six-party process stagnated, North Korea shut down its 5MW reactor in April 2005 and removed the spent fuel. The reactor had been operating since February 2003, meaning that it could have produced enough plutonium for between one and three nuclear devices from its spent fuel. However, it would take a few months for North Korean engineers to extract the plutonium from the spent fuel rods. In July 2005, satellite imagery indicated that the reactor had begun operations once again.

On 19 September 2005, the fourth round of Six-Party Talks concluded and the six parties signed a Statement of Principles, whereby North Korea would abandon its nuclear programs and return to the NPT and the IAEA safeguards regime at "an early date." The United States stated that it had no intention of attacking North Korea with nuclear or conventional weapons, and Washington affirmed that it has no nuclear weapons deployed in South Korea. The parties also agreed that the 1992 Joint Declaration on the Denuclearization of the Korean Peninsula, which prohibited uranium enrichment or plutonium reprocessing, should be observed and implemented.

Although hailed as a breakthrough by some participants, the viability of the Statement of Principles was immediately brought into question by North Korean and U.S. actions. The parties disagreed over the implications of the Statement of Principles for LWR transfer to North Korea. While Pyongyang argued that the six-party statement permitted LWR transfer, Washington countered that this was not guaranteed under the statement and could only occur after North Korea had dismantled its existing nuclear program. Shortly after signing the agreement in Beijing, the U.S. Treasury Department announced that U.S. financial institutions were barred from having correspondent accounts with Banco Delta Asia (BDA), a Macao-based bank, which it accused of assisting North Korea in illicit transactions. North Korea asserted that unless the so-called "sanctions" were lifted, Pyongyang would not carry out its part of the September 2005 agreement. Due to these and other disagreements, the Six-Party Talks stalemated, and the Statement of Principles remained dormant for more than 18 months.

CHAPTER TEN
A NUCLEAR TEST, FAILED NEGOTIATIONS AND ANOTHER NUCLEAR TEST

The nuclear crisis on the Korean Peninsula continued to deteriorate throughout 2006, reaching a low point in October when North Korea conducted its first nuclear test at 10:35AM (local time) at Mount Mantap, Punggye-ri, Gilju-gun, North Hamgyeong Province. The Korean Central News Agency (KCNA) announced that the test was conducted at a "stirring time when all the people of the country are making a great leap forward in the building of a great prosperous powerful socialist nation." The North Korean nuclear test did not, however, produce a significant yield. The yield from this test appeared to be less than 1 kiloton. North Korea was reportedly expecting at least a 4 kiloton yield, possibly indicating that the North Korean plutonium program still had a number of technical hurdles to overcome before it would have a nuclear warhead.

Immediately following the test, UNSC Resolution 1718 imposed sanctions on North Korea. After intense diplomatic activities by the Chinese government and others involved in the Six-Party process, the parties met again in December 2006 following a hiatus of more than a year. However, these talks ended without any sign of progress. In what appeared to be a breakthrough in the negotiations, the six parties in February 2007 agreed on the Initial Actions for the Implementation of the Joint

Statement, whereby North Korea agreed to abandon all of its nuclear weapons and existing nuclear programs, and to return to the NPT and the IAEA safeguards regime, in exchange for a package of incentives that included the provision of energy assistance to North Korea by the other parties. The agreement also established a 60-day deadline during which North Korea was to shut down and seal its main nuclear facilities at Yongbyon-kun under IAEA supervision. Additionally, the United States agreed to release the approximately $25 million in North Korean assets held at the Macao-based Banco Delta Asia. However, the BDA part of the agreement again became a sticking point; much of the international financial community, concerned about the possible legal ramifications of dealing with a bank that was technically still under U.S. sanctions, refused to take part in the transfer of the funds. The issue was eventually resolved when a Russian bank agreed to transfer the funds in June 2007.

After the February 2007 agreement, North Korea extended invitations to IAEA officials, opening the door to reestablishing its relationship with the Agency. In July 2007, North Korea began shutting down and sealing it main nuclear facilities at Yongbyon-kun under IAEA supervision. Further progress was made in the Six-Party Talks when the parties adopted the Second Action Plan, calling on North Korea to disable its main nuclear facilities and submit a complete and correct declaration of all its nuclear programs by 31 December

2007. While disablement activities on North Korea's three key plutonium production facilities at Yongbyon-kun (the 5MW experimental reactor, the Radiochemistry Laboratory and the Fuel Fabrication Plant) progressed, North Korea failed to meet the 31 December deadline to submit its declaration. Sharp disagreements over North Korea's past illicit procurement efforts and controversies surrounding suspected North Korean nuclear cooperation with Syria proved to be the key sticking points.

Almost six months past the deadline, on 26 June 2008, North Korea submitted its much-awaited declaration. While the contents of North Korea's declaration have not been disclosed to the public, various media reports claimed that the declaration failed to address both North Korea's alleged uranium enrichment program and suspicions of its nuclear cooperation with countries such as Syria. Despite problems with the declarations, the Bush administration notified the U.S. Congress that it planned to remove North Korea from the U.S. list of state sponsors of terrorism, and also issued a proclamation lifting some sanctions under the Trading with the Enemy Act. Following the U.S. government's actions, North Korea demolished the cooling tower at the Yongbyon 5MW reactor, an event broadcasted by international media.

Delays with the U.S. removal of North Korea from the state sponsors of terrorism list contributed to North Korean delays in meeting its own commitments, and eventually Pyongyang announced in late August 2008 that it had restored the nuclear facilities in Yongbyon-kun, and barred international inspectors from accessing the site. On 11 October 2008, the United States finally dropped North Korea from the terrorism list after reaching a deal in which North Korea agreed to resume the disabling of its nuclear facilities, and to allow inspectors access to the nuclear sites. The six parties then resumed negotiations to map out a verification plan in Beijing in December 2008. These negotiations focused on ways to verify the disablement of North Korea's nuclear program, including taking nuclear samples. However, the negotiations failed to reach an agreement on a verification protocol, and the issue remains stalled.

After a dispute over rocket launches in March 2009, North Korea kicked out IAEA and U.S. inspectors and began to rebuild the Yongbyon 5MW reactor for the purpose of reprocessing plutonium from its spent fuel rods, in contravention of its previous promises at the Six-Party Talks. On 25 May 2009, North Korea conducted its second nuclear test. KCNA announced that Pyongyang had carried out the nuclear test, and that it "was safely conducted on a new higher level in terms of its explosive power and technology of its control." Initial estimates from the U.S. government showed the test causing seismic activity equivalent to a magnitude of 4.7 on the Richter scale, and located close to the site of the first nuclear test in 2006. Early estimates pointed to a possible yield for the test of between 2 and 8 kilotons, with about 4 kilotons being most likely. The United Nations Security Council released Resolution 1874; in response Pyongyang announced that "the processing of uranium enrichment will be commenced." North Korea further indicated that it did not intend to return to the Six-Party Talks, and asserted that it would not be bound by agreements made earlier through this forum.

Tensions continued to rise in 2010 and 2011. North Korean leader Kim Jong II visited China three times within one year, each time indicating he was willing to proceed with denuclearization efforts; however North Korea also engaged in several military confrontations with the South. In March 2010, North Korea torpedoed a South Korean ship killing 46 sailors, and in November of the same year it shelled Yeonpyeong Island, killing four South Koreans, including two civilians. On 15 March 2011, Pyongyang announced its willingness to return to the Six-Party Talks without preconditions, and agreed to discuss its uranium enrichment program.

However, in March 2010, North Korea announced the construction of a light-water reactor (LWR) at Yongbyon. U.S. nuclear expert Siegfried Hecker confirmed that construction for a 25-30MW experimental LWR had commenced during his November 2010 visit. In November 2011 analysts estimated that the experimental LWR may be externally complete within the next year, but operations are unlikely to begin for another two to three years as machinery and equipment must be loaded and installed. Additionally, Hecker reported that North Korea had completed the construction of a uranium enrichment facility at Yongbyon with 2,000 P-2 centrifuges in six cascades. Although satellite imagery showed that activity had been halted since late April 2014, new imagery from September 2015 show increased activity, likely indicating increased uranium production.

CHAPTER ELEVEN
RECENT DEVELOPMENTS AND CURRENT STATUS

The death of Kim Jong II in December 2011 left much of the world speculating about its impact on North Korea's nuclear program and the Six-Party Talks. After a series of bilateral talks with the U.S., North Korea announced a moratorium on nuclear testing, uranium enrichment and long-range missile tests on 29 February 2012 in exchange for food aid. However, the U.S. withdrew its offer of food aid, after North Korea attempted to launch a satellite into orbit using an Unha rocket on 12 April 2012. The U.S. considers the space launch a violation of the agreement as well as UN Security Council Resolutions 1718 and 1874, because the rocket is not materially different from a long-range ballistic missile. North Korea launched an additional Unha rocket from the Sohae Center in December 2012. This rocket successfully placed a satellite into orbit, and the UN Security Council followed up with Resolution 2087 demanding North Korea end its nuclear and missile programs.

On 12 February 2013, North Korea conducted a third nuclear test at the Punggye-ri Nuclear Test Facility. The USGS reported a 5.1 magnitude seismic shock in the vicinity of the test site. North Korea claimed to have successfully tested a "lighter, miniaturized atomic bomb."

In April 2013, North Korean state media announced that Pyongyang was restarting its 5MW graphite-moderated reactor and uranium enrichment plant at Yongbyon. Though the original cooling tower is now destroyed, satellite analysis confirmed activity consistent with connecting cooling pipes from the 5MW reactor to the adjacent river. By August 2013, satellite imagery confirmed steam venting from the reactor's turbine and generator building. Meanwhile, the external work on the adjacent experimental light water reactor appears to have concluded based on January 2014 satellite imagery; however, it is unlikely the reactor will be fully operational within the next 1-2 years.

Major excavations for water channels and sand dam construction were completed in March 2014 at Yongbyon to supply water to the 5MW reactor's secondary cooling system and the experimental light water reactor's primary cooling system after extensive rainfall and flooding in July 2013 caused damage to the new system. Dam failure could present major safety issues as a reliable supply of water for the secondary cooling system is essential to prevent overheating in the reactor core.

Following a March 2014 KCNA announcement stating the DPRK's intention to conduct a "new form" of nuclear testing, many have speculated about when and how North Korea might conduct another nuclear test. The country may test a new HEU-type device, or alternatively, conduct a series of subterranean tests simultaneously, known as salvo testing. September 2015 commercial satellite imagery indicated increased activity at the Punggye-ri nuclear test site as well as the Yongbyon Nuclear Research Center. The 5 MW reactor and radiochemical laboratory, both key facilities in the production of plutonium, showed high levels of vehicular activity. Similarly, improvements to the Pyongsan mining and milling facility likely indicate an increase in the production of uranium yellowcake.

In December 2015, according to state-run Rodong Sinmun, Kim Jong Un claimed thermonuclear capabilities during his visit to the Pyongchon Revolutionary Site. This claim was met with much skepticism from the international community.

On 06 January, 2016 the DPRK announced it tested its first hydrogen bomb. The CTBTO's seismic monitoring stations detected an event similar to the 2013 test in the Punggye-Ri area, the site of North Korea's three previous tests, leading experts to believe there was a nuclear weapon test. So far, not enough information has come to the surface to confirm whether it was a hydrogen bomb, a boosted fission device or a simple fission device.

CHAPTER TWELVE
HAS NORTH KOREA GOT THE BOMB?

Technically yes, but not the means to deliver it via a missile - yet.

North Korea said it conducted four successful nuclear tests in 2006, 2009, 2013 and 2016.

Analysts believe the first two tests used plutonium, but whether the North used plutonium or uranium as the starting material for the 2013 test is unclear.

While these three were atomic bomb tests, North Korea said its test in January 2016 was of a more powerful hydrogen bomb. Again, the starting material is unclear and experts cast doubt given the size of the explosion registered.

H-bombs use fusion - the merging of atoms - to unleash massive amounts of energy, whereas atomic bombs use nuclear fission, or the splitting of atoms.

Shortly after that test Pyongyang launched a satellite, a launch widely seen as a test of long-range missile technology.

The US said in February it had intelligence indicating that North Korea could soon have enough plutonium for nuclear weapons and was taking steps in making a long-range missile system.

CHAPTER THIRTEEN
WHAT TO KNOW ABOUT NORTH KOREAN NUCLEAR PROGRAM

The Yongbyon site is thought to be its main nuclear facility. The North has pledged several times to halt operations there and even destroyed the cooling tower in 2008 as part of a disarmament-for-aid deal.

However, the US never believed Pyongyang was fully disclosing all of its nuclear facilities - a suspicion bolstered when North Korea unveiled a uranium enrichment facility at Yongbyon, purportedly for electricity generation, to US scientist Siegfried Hecker in 2010.

In March 2013, after a war of words with the US and with new UN sanctions over the North's third nuclear test, Pyongyang vowed to restart all facilities at Yongbyon.

In 2015 a US think tank said satellite pictures suggested the reactor at Yongbyon may have been restarted. Then in September, state media announced that "normal operation" had started at the production plant.

The January 2016 test was said to have been carried out at the Punggye-ri site.

Both the US and South Korea have also said that they believed the North had additional sites linked to a uranium-enrichment program. The country has plentiful reserves of uranium ore.

The United Nations tightened sanctions against North Korea following its third nuclear test

The US, Russia, China, Japan and South Korea have engaged the North in multiple rounds of negotiations known as six-party talks, but none of this has ultimately deterred Pyongyang.

In 2005, North Korea agreed to a landmark deal to give up its nuclear ambitions in return for economic aid and political concessions. But implementing it proved difficult and talks stalled in 2009.

Then in 2012, North Korea suddenly announced it would suspend nuclear activities and place a moratorium on missile tests in exchange for US food aid,. But this came to nothing when Pyongyang tried to launch a rocket in April that year.

The UN further tightened sanctions after the 2013 test.

The 2016 test brought another round of universal international condemnation, including from China, the North's main ally.

North Korea's nuclear actions have become a cause of concern among its neighbors, especially South Korea

After its 2013 test and again in 2015, North Korea claimed it had "miniaturized" a device, or made a device small enough to fit a nuclear warhead onto a missile - which the US cast doubt on.

Pyongyang also said the 2013 test had a much greater yield than the devices detonated in previous tests. It was indeed larger in force than previous ones, but monitors failed to detect radioactive isotopes - a key indicator - so uncertainty remains.

Claims of an underground test of a hydrogen bomb in January 2016 were met with plenty of skepticism.

Initial estimates put the blast in the 10 to 15 kiloton range, whereas a full thermonuclear blast would be closer to 100 kilotons.

North Korea again claimed this was a successful test of a miniaturized device, and again it has not been verified.

References

"North Korea Announces That It Has Detonated First Hydrogen Bomb". New York Times. 5 January 2016.

^ Park, Jeffrey (May 26, 2009). "The North Korean nuclear test: What the seismic data says". "Bulletin of the Atomic Scientists". Retrieved May 28, 2009. External link in |publisher= (help)

^ Geoff Brumfiel (February 3, 2012). "Isotopes hint at North Korean nuclear test". Nature. doi:10.1038/nature.2012.9972.

^ De Geer, Lars-Erik (2012). "Radionuclide Evidence for Low-Yield Nuclear Testing in North Korea in April/May 2010". Science and Global Security 20 (1): 1–29. doi:10.1080/08929882.2012.652558.

^ a b c "North Korea's Estimated Stocks of Plutonium and Weapon-Grade Uranium" (PDF). August 16, 2012. Retrieved March 7, 2013.

^ Stockton, Nick (January 6, 2016). "Science Can Tell if North Korea's Test Was Really an H-Bomb". wired. Retrieved 9 January 2016.

^ a b Military and Security Developments Involving the Democratic People's Republic of Korea (PDF) (Report). U.S. Department of Defense. 2012. Retrieved 23 May 2013.

^ "CNN.com - North Korea leaves nuclear pact - Jan. 10, 2003".

^ Burns, Robert; Gearan, Anne (October 13, 2006). "U.S.: Test Points to N. Korea Nuke Blast". The Washington Post.

^ "North Korea Nuclear Test Confirmed by U.S. Intelligence Agency". Bloomberg. October 16, 2006. Retrieved October 16, 2006.

^ a b North Korea's first nuclear test Yield estimates section

^ a b "Usher in a great heyday of Songun Korea full of confidence in victory". The Pyongyang Times. January 6, 2007. p. 1.

^ Richard Lloyd Parry (April 24, 2009). "North Korea is fully fledged nuclear power, experts agree". The Times (Tokyo) (London). Retrieved December 1, 2010.

^ a b [1] North Korea's Nuclear test Explosion, 2009. SIPRI

^ "North Korea's new nuclear test raises universal condemnation". NPSGlobal Foundation. May 25, 2009. Retrieved December 1, 2010.

^ 2013-02-12 02:57:51 (mb 5.1) NORTH KOREA 41.3 129.1 (4cc01) (Report). USGS. February 11, 2013. Retrieved February 12, 2013.

^ "North Korea appears to conduct 3rd nuclear test, officials and experts say". CNN. February 12, 2013. Retrieved February 12, 2013.

^ Choi He-suk (February 14, 2013). "Estimates differ on size of N.K. blast". The Korea Herald. Retrieved February 17, 2013.

^ "Nuke test air samples are a bust". 15 February 2013. Retrieved 16 February 2013.

^ "How Powerful Was N.Korea's Nuke Test?". The Chosun Ilbo. February 14, 2013. Retrieved February 17, 2013.

^ M5.1 - 21km ENE of Sungjibaegam, North Korea (Report). USGS. January 6, 2016. Retrieved January 6, 2016.

^ "North Korea claims fully successful hydrogen bomb test". Russia Today. January 5, 2016. Retrieved January 5, 2016.

^ no by-line.-->. "N Korean nuclear test condemned as intolerable provocation". Channel News Asia. Mediacorp. Retrieved 6 January 2016.

^ Windrem, Robert. "North Korea Likely Lying About Hydrogen Bomb Test, Experts Say". NBC News. Retrieved 6 January 2016.

^ "North Korea fires long-range rocket despite warnings - BBC News". BBC News. Retrieved 2016-02-08.

^ a b John Pike. "Nuclear Weapons Program". globalsecurity.org.

^ Lee Jae-Bong (December 15, 2008 (Korean) February 17, 2009 (English)). "U.S. Deployment of Nuclear

Weapons in 1950s South Korea & North Korea's Nuclear Development: Toward Denuclearization of the Korean Peninsula (English version)". The Asia-Pacific Journal. Retrieved April 4, 2012. Check date values in: |date= (help)

^ James Clay Moltz and Alexandre Y. Mansourov (eds.): The North Korean Nuclear Program. Routledge, 2000. ISBN 0-415-92369-7

^ "Research Reactor Details – IRT-DPRK". International Atomic Energy Agency. 30 July 1996. Retrieved 14 February 2007.

^ http://www.znf.uni-hamburg.de/MasterSchoeppner.pdf

^ a b c d e f g "Fact Sheet on DPRK Nuclear Safeguards". iaea.org.

^ The U.S.-North Korean Agreed Framework at a Glance, Fact Sheet, Arms Control Association.

^ [2], additional text.

^ "Non-Proliferation Treaty". Dosfan.lib.uic.edu. October 21, 1994. Retrieved March 1, 2012.

^ [3], additional text.

^ Washington Post, "North 'bribed its way to nuclear statehood'", Japan Times, July 8, 2011, p. 4.

^ "North Korea Confirms It Has Nuclear Weapons". Fox News. 11 February 2005. Retrieved 8 March 2013.

^ Traynor, Ian; Watts, Jonathan; Borger, Julian (20 September 2005). "North Korea vows to abandon nuclear weapons project". The Guardian (London). Retrieved 8 March 2013.

^ N. Korea Plans to Shut Down Nuke Facility. Newsmax. March 17, 2007.

^ "UN confirms N Korea nuclear halt". BBC News. July 16, 2007. Retrieved July 16, 2007.

^ a b Steven Lee Myers; Choe Sang-Hun (February 29, 2012). "North Korea Agrees to Curb Nuclear Work; U.S. Offers Aid". The New York Times. Retrieved February 29, 2012.

^ "DPRK Foreign Ministry Spokesman on Result of DPRK-U.S. Talks". Korean Central News Agency. February 29, 2012. Retrieved March 3, 2012.

^ "U.S.-DPRK Bilateral Discussions". U.S. Department of State. February 29, 2012. Retrieved March 3, 2012.

^ "US stops food aid to North Korea after missile launch". Reuters. April 13, 2012.

^ "To React to Nuclear Weapons in Kind Is DPRK's Mode of Counteraction: Rodong Sinmun". Korean Central News Agency. 11 January 2016. Retrieved 11 January 2016.

^ Flower of Reunification (DPRK film [unknown publisher]; official English translation), c. 1989 (Part 3/7)

^ http://fpc.state.gov/documents/organization/71870.pdf

^ http://iis-db.stanford.edu/pubs/22944/Hecker-Lee-Braun_North_Koreas_Choice.pdf

^ a b c "North Korea's nuclear tests". BBC News.

^ Glenn Kessler, Far-Reaching U.S. Plan Impaired N. Korea Deal: Demands Began to Undo Nuclear Accord, The Washington Post, p. A20, September 26, 2008.

^ Demetri Sevastopulo (October 10, 2008). "Bush removes North Korea from terror list". Financial Times. Retrieved October 10, 2008.

^ "N. Korea Says It Has Restarted Nuclear Facilities list". Fox News. Associated Press. April 25, 2009. Retrieved April 25, 2009.

^ Russia Today (April 26, 2009). "North Korea: return of the nukes". RT. Retrieved May 22, 2009.

^ "N. Korea Says It Conducted 2nd Nuclear Test". Fox News. Associated Press. May 25, 2009. Retrieved May 25, 2009.

^ "朝鲜(疑爆)Ms4.9地震" (in Chinese). Archived from the original on August 9, 2014.

^ "Press Release: On the CTBTO's detection in North Korea". CTBTO. Retrieved 12 February 2013.

^ <no by-line.--> (February 12, 2013). "North Korea confirms 'successful' nuclear test". The Telegraph (London, England). Retrieved January 6, 2016.

^ "How Powerful Was N.Korea's Nuke Test?". The Chosun Ilbo. 14 February 2013. Retrieved 17 February 2013.

^ "North Korea has a hydrogen bomb, says Kim Jong-un". The Guardian. Reuters. 10 December 2015. Retrieved 6 January 2016.

^ Sang-hun, Choe (10 December 2015). "Kim Jong-Un's Claim of North Korea Hydrogen Bomb Draws Skepticism". The New York Times. Retrieved 6 January 2016.

^ "News from The Associated Press". hosted.ap.org.

^ Siobhan Fenton (January 6, 2016). "North Korea hydrogen bomb test: Experts cast doubt on country's claims". The Independent.

^ "News from The Associated Press".

^ no by-line.--> (6 January 2016). "North Korea nuclear H-bomb claims met by scepticism". BBC News Asia. BBC. Retrieved 6 January 2016.

^ no by-line.-->. "N Korean nuclear test condemned as intolerable provocation". Channel News Asia. Mediacorp. Retrieved 6 January 2016.

^ "North Korea's Hydrogen Bomb Claim Strains Ties With China". The New York Times. January 7, 2016.

^ Joo, Seung-Hoo (2000). Gorbachev's Foreign Policy Toward the Korean Peninsula, 1985–1991: Power and

Reform. E. Mellen Press. p. 205. ISBN 978-0-7734-7817-6.

Albright, David; Berkhout, Frans; Walker, William (1997). Plutonium and Highly Enriched Uranium, 1996: World Inventories, Capabilities, and Policies. Stockholm International Peace Research Institute. p. 303. ISBN 978-0-19-828009-5.

^ The North Korean Plutonium Stock, February 2007, by David Albright and Paul Brannan, Institute for Science and International Security (ISIS), February 20, 2007.

^ Albright, David; Brannan, Paul (June 26, 2006)

^ "Weapons of Mass Destruction". Retrieved November 6, 2012.

^ "International Atomic Energy Agency (IAEA).". International Atomic Energy Agency. Retrieved November 5, 2012.

^ Busch, Nathan E. (2004). No End in Sight: The Continuing Menace of Nuclear Proliferation. University Press of Kentucky. p. 251. ISBN 978-0-8131-2323-3.

^ Siegfried S. Hecker (12 May 2009). "The risks of North Korea's nuclear restart". Bulletin of the Atomic Scientists. Retrieved 5 November 2009.

^ "FSI - CISAC - North Korea's Choice: Bombs Over Electricity". stanford.edu.

^ "Defiant NKorea resumes nuclear program". Brisbane Times. 25 April 2014. Retrieved 16 October 2014.

^ Bodansky, Yossef; Forrest, Vaughn S. (August 11, 1994). Pyongyang and the US nuclear gambit. Congressional Documents. GlobalSecurity.org.

^ 북한내 핵실험 가능 추정지역 최소 8곳 [Minimum of eight nuclear test in North Korea can be estimated]. BreakNews (in Korean). October 6, 2006. Retrieved November 1, 2009.

^ Nuclear Weapons Program – North Korea History section paragraph 1. Federation of American Scientists. Accessed 5 April 2013.

^ "Khan 'gave N Korea centrifuges'". BBC News. August 24, 2005. Retrieved March 1, 2012.

^ "ABC News: ABC Exclusive: Pakistani Bomb Scientist Breaks Silence". ABC News. ABC News (USA). May 30, 2008. Retrieved March 1, 2012.

^ "N Korea 'admits nuclear programme". BBC News. October 17, 2002. Retrieved October 5, 2006.

^ Sanger, David E.; Broad, William J. (March 1, 2007). "U.S. Had Doubts on North Korean Uranium Drive". The New York Times. Retrieved March 1, 2007.

^ Kessler, Glenn (March 1, 2007). "New Doubts on Nuclear Efforts by North Korea". The Washington Post. Retrieved March 1, 2007.

^ "Nuclear weapons: Who has what?". CNN. Retrieved April 16, 2013.

^ Daily chart: Mutually assured ambiguity. The Economist (2013-06-03). Retrieved on 2013-07-12.

^ a b "Arms Control and Proliferation Profile: North Korea". Arms Control Association. April 2013. Retrieved 9 August 2013.

^ "North Korean Military Capabilities". Archived from the original on September 11, 2006. Retrieved October 5, 2006.

^ a b c "DPRK – Chemical Weapons Program". GlobalSecurity. 2003. Retrieved 9 August 2013.

^ Jon Herskovitz (18 June 2009). "North Korea chemical weapons threaten region: report". Reuters. Retrieved 9 August 2013.

^ N. Korea threat beyond neighbor, Omaha World-Herald, 28 November 2010

^ Ryall, Julian (4 July 2015). "Defector says North Korea tests germ warfare on disabled". The Age. Retrieved 11 January 2016.

^ Asher-Schaprio, Avi (9 July 2015). "Did North Korea Really Publish Pictures of a Biological Weapons Facility?". VICE News. Retrieved 11 January 2016.

^ "Kim Jong-un invites entire US government to visit pesticide plant". The Daily Telegraph. 15 July 2015. Retrieved 11 January 2016.

^ a b c d e f Markus Schiller (2012). Characterizing the North Korean Nuclear Missile Threat (Report). RAND Corporation. ISBN 978-0-8330-7621-2. TR-1268-TSF. Retrieved January 19, 2013.

^ Ravi Shekhar Narain Singh (2005). Asian Strategic And Military Perspective. Lancer Publishers. ISBN 817062245X.

^ a b John Pomfret and Walter Pincus (December 1, 2010). "Experts question North Korea-Iran missile link from WikiLeaks document release". The Washington Post. Retrieved June 13, 2012.

^ http://www.isis-online.org/publications/dprk/DPRKplutoniumFEB.pdf

^ "North Korea warned about missile". BBC News. June 18, 2006. Retrieved October 5, 2006.

^ "Nikkei Interview Article Computer Translation". Excite-webtl.jp. Retrieved March 1, 2012.

^ "The North Korean Rocket Launch: International Reactions and Implications". James Martin Center for Nonproliferation Studies. 27 April 2009. Retrieved 6 April 2014.

^ a b "North Korea missile tests defy UN". BBC News. July 4, 2009. Retrieved July 2, 2009.

^ Choe Sang-Hon (July 2, 2009). "North Korea Test-Fires 4 Short-Range Missiles". The New York Times. Retrieved July 2, 2009.

^ a b Steve Chao (July 2, 2009). "North Korea fires series of missiles". Al Jazeera. Retrieved July 2, 2009.

^ BBC News - How potent are North Korea's threats?

^ SHANKER, THOM (April 11, 2013). "Pentagon Says Nuclear Missile Is in Grasp for North Korea". The New York Times. Retrieved 11 April 2013.

^ "Could North Korean Missiles Hit the U.S.?". Retrieved 25 March 2013.

^ Eric Talmadge (April 26, 2012). "Analysts say North Korea's new missiles are fakes". The Independent (London). Retrieved April 29, 2012.

^ John Pike. "Rodong-1". Globalsecurity.org. Retrieved March 1, 2012.

^ "CRS report for Congress" (PDF). Retrieved March 1, 2012.

^ Pekdosan-1 ("Taepodong-1"), skyrocket.de

^ E:\PICKUP\89797A

^ Ballistic and Cruise Missile Threat (PDF). National Air and Space Intelligence Center (Report) (Air Force Intelligence, Surveillance and Reconnaissance Agency). April 2009. NASIC-1031-0985-09. Retrieved 20 February 2013.

^ http://38north.org/2015/03/jschilling031215/

^ a b c
http://www.defense.gov/Portals/1/Documents/pubs/North_Korea_Military_Power_Report_2013-2014.pdf

^ "Facts about North Korea's Musudan missile". AFP (GlobalPost). 8 April 2013. Retrieved 10 April 2013. IHS Jane's puts the estimated range at anywhere between 2,500 and 4,000 kilometres ... potential payload size has been put at 1.0-1.25 tonnes.

^ "komid-un". CNN. April 25, 2009. Retrieved March 1, 2012.

^ KOMID Overseas at the Wayback Machine (archived August 29, 2010)[dead link]

^ KOMID and Iran at the Wayback Machine (archived October 11, 2010)[dead link]

^ KOMID and Taiwan at the Wayback Machine (archived November 19, 2010)[dead link]

^ "KOMID and Syria". Breitbart.com. Archived from the original on January 19, 2012. Retrieved March 1, 2012.

^ KOMID's $100 million sales Archived May 12, 2013 at the Wayback Machine

^ "Hap Heng in Macau". CNN. Retrieved March 1, 2012.

^ 印務局 Internet Team. "Ko Myong Hun and Kim Song In". Bo.io.gov.mo. Retrieved March 1, 2012.

^ "国家简介-国际-朝鲜民主主义人民共和国大使馆". People's Daily. Retrieved March 1, 2012.

^ "Ko and Komid". China Daily. Retrieved March 1, 2012.

^ McElroy, Damien (November 12, 2010). "North Korea 'runs international nuclear smuggling network'". The Daily Telegraph (UK). Retrieved March 1, 2012.

^ "BBC NEWS - South Asia - Powell says nuclear ring broken". bbc.co.uk.

^ "Report to Congress, January – June 1999. Unclassified Report to Congress on the Acquisition of Technology Relating to Weapons of Mass Destruction and Advanced Conventional Munitions. Central Intelligence Agency". Cia.gov. Retrieved March 1, 2012.

^ William J. Broad; James Glanz; David E. Sanger (28 November 2010). "Iran Fortifies Its Arsenal With the Aid of North Korea". The New York Times. Archived from the original on 29 November 2010. Retrieved 28 November 2010.

^ "Syria - Country Profiles - NTI". NTI: Nuclear Threat Initiative.

^ Bermudez, Joseph S. (1999). "A History of Ballistic Missile Development in the DPRK: First Ballistic Missiles, 1979–1989". James Martin Center for Nonproliferation Studies. Retrieved February 14, 2008.

^ "IISS report". Iiss.org. Retrieved March 1, 2012.

^ North Korea Missile Milestones – 1969–2005 Archived December 4, 2010 at the Wayback Machine